Fire
In
Focus

An Action
Portfolio

Fire In Focus

An Action Portfolio

Photographs by
Thomas K. Wanstall

Foreword by
John D. Peige

Design by
Jon M. Nelson

SQUAREBOOKS
MILL VALLEY, CALIFORNIA

A Note from the Photographer

I could complain about the trials and tribulations, the rigors of long hours and inclement weather, and the many personal sacrifices that went into the creation of this book. But I won't. Because, to tell the truth, I have thoroughly enjoyed every minute of every hour that I spent taking these photographs. Sure there were many freezing nights in winter and sweltering afternoons in summer. Yet as cold as I was on those winter nights, the firefighters were colder. And as uncomfortable as I was in the damp summer heat, the firefighters were falling from heat exhaustion.

Fascination with fire engines and firefighting has been with me for years. No matter where I am or what I may be doing, the sound of sirens and air horns still diverts my attention. My admiration for firefighters and the tasks they perform is unbounded. Their ability to perform heroically as individuals or as a team is limitless.

For a photographer the scene of a fire can be an oasis of picture possibilities. The entire spectrum of human experience and emotion is present: fear, panic, pain, tragedy, empathy, compassion, valor and, at times, even humor.

Several years ago, as I sat looking through stacks of fire photos, I became concerned that they were all starting to look alike. How, I wondered, could so many incidents and so many places and so many times appear so similar? My photographs were not communicating what I had felt at the fire scene.

Upon reflection, I began to understand that there was more to taking great fire photos than just being there and snapping the shutter. I learned it was necessary to spend more time observing what was going on *around* me, not just in front of me. My camera would record the moments—I just had to recognize them. When I trained myself to observe, to recognize and to record, in that order and in a systematic fashion, the results were significantly more satisfying.

The question "Howja do it?" arises so often that to satisfy the

Published by Baron Wolman/SQUAREBOOKS

Copyright © 1984 by Squarebooks, Inc.

Photographs copyright © 1984 by Thomas K. Wanstall

Color prints are available from the photographer. For information write Thomas K. Wanstall, Post Office Box 17, Yonkers, NY 10703.

ISBN: 0-916290-21-2

SQUAREBOOKS, Inc.
Post Office Box 1000
Mill Valley, CA 94942

Typesetting: Mark Decker

Printed in Japan by Dai Nippon Printing Company, Ltd., Tokyo

curious I have diagrammed and explained in detail a couple of the more popular photographs. The photo on page *viii* required a fairly extensive setup, but it can be done by anyone willing to give it a try. The shot on page 3 necessitated a good working knowledge of camera and film, a quick evaluation of the available light and fast composition . . . plus a bit of just plain luck, of course.

I try to keep my equipment to a manageable minimum, carrying enough to get the job done without needing a pack mule for an assistant. I am partial to my Contax cameras and Zeiss lenses. My camera bag usually contains two Contax RTS-II bodies with 15mm, 25mm, 45mm, 50mm, 85mm and 135mm Zeiss lenses. I carry a Yashica ML 80-200mm zoom lens, but seldom use it, as I prefer fixed focus lenses. I also use two Contax TLA-30 dedicated strobes and two Vivitar 283s, all powered by Quantum battery packs. Small yet powerful, these lightweight flash units are gems.

I cannot say enough about the need for adequate and reliable scanning monitors—without them the task of taking good fire photos would be impossible. The scanner is as important as the camera itself, for if I am to make the best possible photos, I must get to the fire quickly, and only scanners can be counted on to alert me and sound the call to action. After all, a fire scene is not like a studio where you can re-shoot if the first-time results are unsatisfactory.

In my car and at home I use a Bearcat 260. It has been engineered specifically for mobile use, yet it is cleverly contoured for home use as well. Neatly compact, its fully illuminated control panel faces upward for full viewing and is within easy reach to facilitate making entries while on the move. To confirm successful entries, an audible tone sounds after each operation key is pushed. The 260 comes fully wired for automobile operation and features a priority channel, delay, search, lockout and even a weather key for instant weather reports outside of the 16 operational channels.

My handheld scanning duties are shared by a well-worn Bearcat Four-six and a new Bearcat 100 programmable, the latter of which is great when I'm traveling around and playing the role of the "visiting fireman."

No project of this scope can be accomplished without the support and cooperation of people. My hand goes out to all the firefighters who perform their duties so well and who are so patient with the ever-present army of photographers that gathers on the scene. My appreciation to the firefighters of the FDNY for their hospitality and cooperation. And my special thanks to all the members of Yonkers Firefighter Local 628 and the Uniformed Fire Officers Association for their many years of extraordinary friendship and assistance in all my photographic endeavors.

To my family, who rallied during the hard times with loving confidence and unquestioning support, my eternal love and gratitude.

"By 1900 the man in the street had forgotten all that his ancestors ever knew about what goes on behind the camouflage of smoke...

If only there were some way of taking the entire public down through a hot, smoke-choked hallway behind a crew of firemen...It might arouse ...some remembrance of the trickery and treachery of fire, and some respect for the men who fight it."

—JOHN V. MORRIS
Fires and Firefighting

Foreword

*E*ver since man first discovered fire, he has felt compelled to preserve its flickering images. From primitive etchings in prehistoric caves, to the exactness of Egyptian hieroglyphics, to the painstaking detail displayed in works by the great masters, man's fascination with recording the fire scene has been universal.

Through the ages, the art of capturing the excitement of roaring flames visibly improved but was still an imperfect science, prejudiced by subjective artistic interpretation. Often the result was a kaleidoscope of impressions condensed into one massive work. The true essence of the fireground—the unrelenting, personal battle of man against fire—was yet to be seen.

Enter the twentieth century, and a whole new way of exposing this age-old struggle comes into focus. The advent of fire photography gives birth to a brave new breed plying the nation's fireground . . . the fire photographer.

And that's what *Fire in Focus* is all about. This all-out portfolio graphically portrays the realities of modern-day firefighting as shot by Tom Wanstall, one of America's pre-mier fire photographers. Capturing much more than just what's burning in front of the lens, Wanstall's work is alive with the hectic and often heroic activity of the moment.

Nowhere is the action more intense, more fleeting, than at the scene of a working fire. Here, thousands of images flash past—the operations of the first company, the placement of men and equipment used in fire attack, the frantic search for and rescue of trapped victims, and the grim aftermath of salvaging what's left after a fire has wreaked havoc—and it's the measure of a Master Fire Photographer (M.F.P. is the highest distinction granted by the International Association of Fire Photographers) to anticipate the right moment to click the shutter. From this supercharged atmosphere, Wanstall's camera tells an unforgettable story of what it's like to fight fire in some of America's toughest neighborhoods. His photos frame the entire spectrum of human experience and emotion: fear, pain, tragedy, courage, compassión, victory, and even, at times, humor.

But what about the man himself?

Wanstall began his fire photography career in his hometown of Yonkers, New York, a city of 200,000 which borders New York City's borough of the Bronx. Many of the same socioeconomic conditions which spawned the infamous "fire problem" of the South Bronx also put Yonkers to the torch. As the fire load increased, so did Wanstall's experience and skills in fire photography. Unlike the conditions found in his 22-year career as a professional photographer for a major airline, the fireground scenes unfolding around him could never be repeated. There were no second or third takes, no studio to manufacture "reality."

As Wanstall worked to gain familiarity with this stark, new environment, his efforts began to yield results. At the same time, his view of the fireground expanded to include the entire New York metropolitan area—from the high-rises of Manhattan, to the tenements of Harlem, to the suburban towns and villages that encircle the world's largest city.

This fireshooter also learned what to look for during the first critical minutes of a working fire. Once at the scene, he

uses several criteria in selecting subjects. He avoids the obvious, looking instead for rescue, high-impact action, good PR for the department or a combination of these elements with an eye to artistic value.

Wanstall feels a responsibility to uphold and promote the image of the firefighter in the community, and spends much of his time in public education efforts. Through slide presentations at various civic organizations, he illustrates just how dangerous the firefighter's job can be.

The general public has little chance to observe what goes on behind the camouflage of smoke and flame. Too often they picture only the thrill and excitement of speeding apparatus, flashing lights and wailing sirens—and their impression of the fire service stops there. It's only through the eyes of photographers like Wanstall that the true grit of the firefighter's job can be revealed.

But Wanstall's interest in the fire service transcends the world of fire photography, and nowhere is it more evident

than in the place where he lives and now works. From the first time you set foot in his home, you know that FIRE is spoken here. His living space is shared by a virtual mini-museum of firematic collectibles surrounded by a preponderance of apparatus models which occupy almost every conceivable flat surface. The fleet comes in all shapes and sizes, and many are collector's items built to exacting detail.

But it's the vertical space that dominates the visitor's eye with a gallery of poster-sized prints that encapsulate the photographer's two very separate job experiences. It is indeed an intriguing juxtaposition to see one wall aglow with the raging hell fires of the urban ghetto and another radiant with scenes from some of the world's most exotic travel locations. The harsh, grim realities of all hope lost confront the subtleties of wild flowers blooming in a paradise lost. Both demand attention, both capture the imagination, both convey a message . . . and perhaps Wanstall's personal vision of life itself lies in the contrast.

JOHN D. PEIGE
Hastings-on-Hudson, N.Y.
April 14, 1984

The
Size-Up

"Organized panic."

That's how one seasoned fire officer describes the conditions unfolding at a working fire, moments after first-due units arrive at the scene. One never knows what to expect. Flames can be shooting out of every window of a building, while the street below is deserted, eerily still. Or, an almost overwhelming scene can be awaiting the first-in rig, as hysterical residents scream with anguish.

To America's Bravest, however, the job is the same—the fire must be fought; the battle, joined.

Experienced firefighters will tell you that no two buildings burn exactly alike. Each must be sized up individually if a good stop is to be achieved. What's a good stop? Basically, confining the fire to where you found it . . . the room, floor, building or, in some cases, block of origin.

Fully involved
Fire extension in this three-story apartment building makes rapid deployment of a ladder pipe necessary. Use of an exterior master stream will deliver maximum water flow in an all-out effort to achieve a rapid knockdown.

One in a million
A strikingly eerie scene is created when a 13,000-volt electrical transformer explodes after being exposed to intense radiant heat from a row of burning frame dwellings.

Working fire
Fully involved on arrival, this fire requires the command chief to communicate instantly with arriving units.

4

Tough going
A battalion chief evaluates the progress of a cellar fire, a notoriously hazardous type of blaze.

Getting the word
Information is passed to the line officer for deployment of arriving companies.

Chief's nightmare

Heavy smoke is showing in an occupied frame building with an unknown life hazard and a serious exposure problem.

Cutting a hole in the roof

(*Opposite*) A must at this taxpayer fire. A taxpayer is a row of stores with a common attic space. If proper ventilation isn't performed quickly to check flame spread, the whole strip could be lost.

tway
HATTERS
HOE REBUILDERS
& HAND BAGS REPAIRED
EDIC SHOE WORK
GOODS REPAIRED
Mazzo
Floor
RUGS
TILE
ICE SKATES
WE HAVE A LARGE
WALLPAPER
Yellow Pages

The pressure's on
The pump panel is the best
place to gauge the progress of
water-supply operations.

Park Place
(Right and Opposite) Placement of apparatus in the early minutes is crucial to effective fireground operations. First-in engine companies must leave room for ladder trucks to set up in front of the fire building. If the job looks like it's going to extra alarms, access should be left so later arriving companies can take effective positions.

It can get crowded
(Left) One advantage of operating with the world's largest fire department is that an almost unlimited amount of men and equipment is available. This massive array of FDNY apparatus turned out for a rapidly spreading four-alarm fire in Brooklyn.

Safety first
Another advantage of a large department is that chief officers can be assigned specialized tasks. Here, the FDNY's safety chief checks roof conditions.

Transmit the second

Command and control at multiple alarms is essential. Often, night operations combined with dismal weather blur the identity of numerous apparatus at the scene. Communications is key for a chief operating at a field command post, where updated status reports and other vital fireground information can be relayed instantly.

The
Attack

*T*he goals in fire attack are invariably the same: saving lives and protecting property. Search and rescue is always key to initial operations, and once it is completed, fire attack begins in earnest.

In this way the fireground resembles the battleground. The difference, though, is that here there is no time to carefully lay plans or marshal troops. In this war zone, action is immediate. Tactics must be implemented at once . . . a tough task when dozens of variables have to be carefully weighed in an unrelenting, adrenalin-charged atmosphere. Results are also immediate. If the strategies are successful, the fire goes out. If not, the long day stretches into an endless night.

Opening up
Firefighters place ladder pipe in service to halt the spread of a rapidly advancing cockloft fire. A cockloft is the concealed space between the top-floor ceiling and the roof of a building.

Chasing the red devil

Sometimes a fire takes a curious route as it spreads through a building. It can race undetected from foundation to attic inside exterior walls built without fire-stops. This is known as balloon construction and has been the death knell for hundreds of wood-frame buildings.

Outside looking in
(Opposite) The most effective way to fight a fire is from the inside. Sometimes, though, it just isn't possible to enter a burning building through the front door, and firefighters are forced to seek an alternative means of entry.

Helping hand
All alone with a charged line, up against the forceful back pressure created by hundreds of gallons of rushing water, a firefighter appreciates getting a hand from a friend.

**Rising to meet
the challenge**

Against all odds
The enemy can be
overwhelming at times.

Blow out
Take an arsonist with a gallon of gasoline and the results are predictable: a heavily involved store fire.

Missing Member
Fellow firefighters rush to render assistance when a crew member suddenly disappears from view in this heavily smoke-charged building. Their fears are short-lived as he reappears unscathed.

Heavy artillery
FDNY's Satellite 2 is in action at a five-alarm blaze in the Bronx. One of five such units in the city, it features an Intelligiant deck gun that can deliver 4700 gallons of water per minute.

Readying the pipe
Operating with rapidly deteriorating fire conditions, crew members man the ladder pipe.

Bridging the gap
In the tight confines of urban areas, flames can readily leap from one building to another. Keeping the blaze out of adjoining structures is known in fire jargon as "protecting the exposures." Sometimes they're protected, sometimes they're not. And when they're not, the battle begins anew.

Shooting the moon

A stubborn fire in the dead of night can make weary firefighters wish for a different kind of moonshine.

Heavily engaged

When all hands are hard at work and each piece of apparatus is committed, it's a good bet that additional alarms will be called, as was the case at this three-bagger.

At death's door
Firefighters search frantically to reach trapped victims . . . only to lose one of their own in this deadly, towering inferno.

Water, water everywhere
(Opposite) Coast Guard fireboat presses its attack at the scene of a general-alarm fire.

Sharpshooter
With the precision of a marksman, a nozzleman zeros in on a column of advancing flames.

Fire for all
(Opposite)
No question about the attraction at this street scene.

Towering above it all

The 75-foot tower ladder has become an important element in the modern fire attack. Capable of delivering 1000 gallons of water per minute into burning buildings, it has a tremendous reach and maneuverability that enable firefighters to direct a stream onto the base of the fire—whether it be in the side yard, on the top floor or on the roof of the structure. The tower ladder can carry up to 1000 pounds in the bucket and is invaluable in removing trapped occupants, since it can remove them to safety without exposing them to the danger of climbing down ladders.

Finding a proper place

Placement of a tower ladder at the fire scene is critical to its successful operation. Not only must the rig be placed directly in front of the fire building, but room must be left to extend the outriggers and four jacks which lift the entire vehicle off the ground. Since it is not equipped with a pump, the tower ladder must be supplied with water by an engine company.

The old-fashioned way

Modern machines have done much for battling the big blazes, but in the final count it is still men with hoses that fight the first and last flames.

From above and below
Deck gun and tower ladder streams combine for a heavy-duty attack, as fire races through a four-story structure.

Humping hose

It's not all glamour and excitement. When the chief calls for more line, it takes a strong back to fill his request.

Face to face

Flames rush out of a burning frame house to meet arriving firefighters.

From all angles

Firefighters must be prepared to do combat on the enemy's ground.
(Opposite) Firefighter prepares to go underground to battle subterranean fire.
(Above) Settling in for a long ground-level operation, firefighters man deck
gun. *(Right)* Aerial attack is made on top-floor fire by twin FDNY tower ladders.

At the high point of his career

(Opposite) Firefighter on ladder pipe attempts to stay on top of things during early-morning blaze.

The aftermath

The remains of a once-proud apartment building stand as mute testimony to a fire's final fury.

Hanging out to dry

An unusual scene is presented after a fire in an industrial paint shop. As firefighters make certain fire is extinguished, freshly painted materials hang to dry in the foreground.

Rescue

*R*escue—the ultimate challenge in fire combat. Victims trapped in a fire building often are confused, disoriented, incoherent or, worse, have been rendered unconscious by the ravaging effects of searing heat and smoke.

Some of the toughest, most dramatic rescues—those that make the front pages and win awards for valor—are made by aerial ladder. This type of rescue is rare, however—attempted only if all other means of escape are blocked.

Unlike an athlete who can fully prepare for the main event, a firefighter never knows when his time has arrived. Rescues happen suddenly, at the worst times and in the worst places. Resolute action is the name of the game.

Hotel horror

An early-morning hotel fire on a bitter cold December day adds up to horror both for trapped residents and for Yonkers, New York, firefighters. As the blaze burns through the roof, firefighters begin initial search. A woman appears in a smoke-filled window at the far right, then suddenly vanishes.

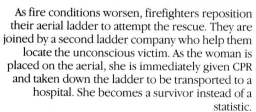

As fire conditions worsen, firefighters reposition their aerial ladder to attempt the rescue. They are joined by a second ladder company who help them locate the unconscious victim. As the woman is placed on the aerial, she is immediately given CPR and taken down the ladder to be transported to a hospital. She becomes a survivor instead of a statistic.

Small fire, big rescue

A woman and a child are trapped in their bedroom when fire in an adjoining room blocks their escape. Firefighters raise a ground ladder to the third-floor window and effect the rescue of the two frightened but otherwise unharmed civilians.

One for the road

Highway accidents often present bizarre problems for firefighters to solve. A heavy crane has fallen from a flatbed trailer, trapping the driver in the car between the machine and the guardrail. Firefighters must cut through the rail to gain access to the trapped victim. Miraculously, the driver sustains relatively minor injuries.

Fear more than fire
Occasionally, occupants are struck with a fear that is far greater than that which befits the existing fire. This woman was unable to exit via the interior stairs and had to be removed by ladder.

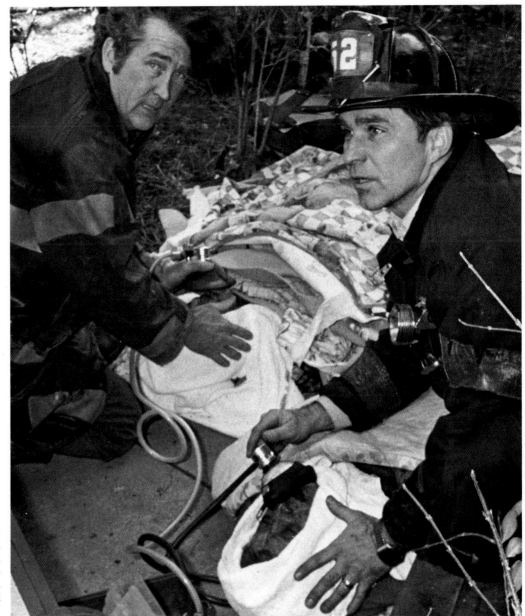

Care and compassion
Flash fires can strike with deadly speed. In addition to battling blazes, firefighters must have the ability to aid the victims with care and compassion.

Hell in the high-rise

The proliferation of high-rise dwellings has presented firefighters with many new and hazardous problems, such as occupants rushing to terraces instead of fire stairs, and oven-like heat within the fire-swept walls. Beleaguered Yonkers firefighters made more than twenty aerial rescues at this general-alarm fire. Many of the occupants were senior citizens and required extra care descending the ladders. Exhaustion became the enemy for many of the embattled warriors.

Midnight arson

Cloaked by darkness, an arsonist has touched off an inferno in a wood-frame structure. Two panic-stricken women, trapped in their third-floor apartment, desperately scream for help. Two police officers arrive on the scene first and attempt to reach them with a painter's ladder that proves to be too short. Their presence, however, keeps the victims from jumping. The first-due engine company arrives not a minute too soon and uses a ground ladder to pluck the women from their plight.

A Fallen Comrade

"I want to fill my calling, Lord,
To give the best in me.
To guard my every neighbor and
Protect his property.
And if according to my fate,
I am to lose my life,
Please bless with your protecting hand
My children and my wife."

from the "Fireman's Prayer"
author unknown

A Gallery of Images

*T*here's more to photographing the fire service than shooting burning buildings and firefighters stretching hose-lines. One needs only to look beyond the fireground to find scenes depicting the everyday life experiences of firefighters at work and at leisure: from the dirt, sweat and tears, to the firehouse humor, to the men and women hard at play. There is the polishing and care of antique apparatus, the strenuous competition between firefighters at the annual musters. These all make up the modern lifestyle of America's Bravest.

Taking a blow
The story of the firefighting experience is often told in the faces of those who perform America's most hazardous calling. There one can see the obvious—the heat, the smoke and the dangers of operating within the confines of the enemy.

Blowing hot and cold

There are those parts of the job which are not so obvious, but which are ever present: the exhaustion in the summer heat, the indescribable discomfort of the winter cold, the proud confidence of knowing you did your job well, the intensity of making the correct command decisions.

Flat out
Firefighters reflect on the ordeal moments after they know they have a "handle" on the fire. One can only speculate on the myriad of thoughts passing through their minds.

Perhaps a miracle
This bible was the sole survivor of a fire that totally destroyed the church in which the book was housed.

Come and get it
A popular person in the firehouse, the chef is often adorned with gifts that are sifted from the rubble of a fire. This chef was presented with a hat to assist him with his cooking duties.

Sixty emotions per hour
From urgency to affection and compassion—few occupations allow one to experience such extremes of feelings in such a short time.

Keeping in touch
Firefighters are really residents of two neighborhoods—the one in which they live and the one in which they work. They're very much a part of their adopted community—while the firehouse is their "office," it is also their home away from home.

A matter of priorities

Firefighters' meals are often interrupted. A two-alarm response halted this meal before it even reached the firehouse.

Recycling

Eastchester, New York, firefighters remove bicycle from the path of oncoming overhaulers. It was saved to be ridden another day.

Muster madness

(Opposite) Firefighters vie for honors in an intense muster competition. *(Above)* Ben Franklin Bridge is the backdrop for Chews Landing snorkel at a muster in Philadelphia. *(Above right)* Valhalla is the place for "old-timers" to show their stuff. *(Bottom right)* As if looming out of the past, this magnificent Seagrave pumper makes its annual appearance at the Croton muster.

Pumping power
Antique rigs demonstrate that they still have what it takes, in the reflection pool at Valhalla.

Top hats
Polished and placed with perfection, these helmets await the judges' scrutiny for top honors in the antique apparatus competition.

Really stacked

Steam pumpers are a treasured tradition of the early days of firefighting. The clattering hoofbeats and the puffing steam were music to the ears of early "buffs," much as the siren and the air horn are to buffs today.

Seeing red

In its day, this rig cleared many roads for the Niagara Hose Company in Burlington, New Jersey, with its spectacular Roto-Ray and glimmering bell.

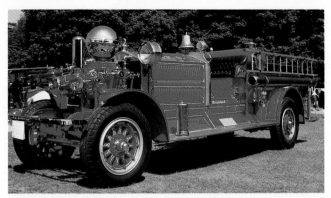

The Fox
No other fire engine has been revered as much as the classic Ahrens Fox, with its distinctive domed centrifugal pump.

Reflections
Thoughts of fighting fires in days gone by.

Imagination
Thoughts of fighting fires in days yet to come.

Questions & Answers

What kind of cameras are best for covering fire?

Single-lens reflex 35mm cameras are best for fire work; they are rugged, flexible and easy to handle, and most brands offer high-tech electronic controls, interchangeable lenses and countless other accessories. Forget about larger format cameras (6x6cm or 6x7cm). They are big, delicate and slow-working, giving only ten to twelve shots per roll of film. The big negatives are nice for making prints, but their ponderous handling characteristics will cause you to miss fast-breaking photo opportunities. I have great admiration for the previous generation of fire and news photographers who got more than their share of wonderful shots with such bulky and hard-to-use equipment.

What brand of 35mm should you buy? I use the superb Contax system, but there are ten or fifteen others from Europe and Japan, which, frankly, can take equally good pictures. Buy the system which offers you flexibility and value. Since there is little interchangeability between brands, consider the system that your friends and colleagues use so you can swap and share various components.

Which lenses should I use?

Most 35mm systems offer a range of their own lenses, from ultra-wide fish-eyes to telephotos a yard long. In addition there are high-quality "off brands," such as Vivitar, Tamron and Kiron, which offer lenses with universal mounts that allow them to be used on virtually any single-lens reflex (SLR). You can certainly take great fire shots with a so-called "normal" 50mm lens, but it is probably the lens I use the least. I suggest adding at least two lenses: a wide-angle (28mm or 35mm) and a telephoto (in the 135mm range). A telephoto zoom, such as a 70mm-210mm, is an alternative, but the zooms tend to have smaller maximum openings, making them more difficult to use in low-light situations. I seldom use my zoom. If I had to, I could get along very well with a three-lens setup (wide, normal, télé), but I go a couple of steps further and always carry an ultra-wide (20mm or 25mm) and a stronger telephoto (180mm, 200mm or 250mm) for those really dramatic shots.

Every photographer has his favorite combination of lenses, so there is no single answer to this question. It's best to start with the basics and then try other focal lengths that match your personal vision. I will say this: The occasional genius can come up with sensational shots using nothing but one camera and a normal 50mm lens. Not me. I rely heavily upon a range of equipment—camera bodies, different lenses, motor drives, strobe lights, the works—to help me get my best fire photos.

Are automatic cameras good for fire action photography?

Automatic (electronic exposure) cameras are fine, but it is very important to understand both their limitations and their wonderful capabilites. Automatic cameras are designed to come up with the optimum exposure in *most-normal situations,* and this they do well. But their circuitry can be fooled, usually by large variations of contrast within a single frame. The classic example is the skier standing in the snow (or the firefighter in front of the huge blaze!) where the automatic metering system is so overwhelmed by the bright background that it badly underexposes the figure in the foreground.

There is still no substitute for actually knowing the settings for those situations you encounter most often. And keep your eye on the automatic exposure indicators to see if the camera is being led astray by the subject. My automatic cameras usually make the correct setting, though I occasionally switch to manual when I sense they are being fooled. If the photo is potentially a great one, I take no chances, and "bracket" the shot by quickly shooting it both ways: the camera's automatic solution and my own.

A word of caution. The earlier half-automatic, half-manual cameras are being replaced by newer ones with totally electronic innards. These cameras often won't work at all if their batteries die.

Some, such as the Contax and Nikon lines of automatics, provide a single manual shutter speed—usually 1/60th or 1/90th of a second—that kicks in if the batteries conk out so that the camera will remain marginally useful without any electronic power. Always keep spare batteries on hand, know how to change them quickly (and in the dark) and know what your cameras can and can't do without full power.

It is sadly ironic that many photo enthusiasts buy a library of "how to" photo books to go along with their expensive equipment purchases, yet they seldom read the little free instruction manual that is packed with every piece of gear. Make sure you know *everything* about your electronic camera. If it starts an argument with you at a free-for-all fire scene, that's hardly the time to drag out a book and try to figure out what has gone wrong! Do not even think of following a fire until you know the function of *every single* button, knob and contol lever on your camera.

Will water damage my equipment?

If you get close to working fire scenes, you and your gear are going to get wet. Count on it. Especially if you get to shoot action, training or evidence inside a fire-struck building. Water will assault you from all sides: misaimed hose streams, spurting leaks in lines, "rain" from waterlogged upper floors, condensing steam. There are several ways to deal with the problem, but unfortunately, none are particularly satisfactory.

One solution is to choose gear designed for underwater work, such as the Nikonos camera and its companion flash units. Also on the market are some excellent low-priced, water-resistant sports cameras, such as the Fujica HD-S. With these, however, you sacrifice flexibility and ease of operation for moisture resistance.

Another protection technique is to wrap cameras and strobes in clear plastic, with only the lens (protected by a clear or skylight filter) protruding. This idea works better with strobes than with camera bodies which constantly have to be adjusted and reloaded.

Since I usually wade right in and get wet, I use waterproof cameras and I also sometimes wrap the Contaxes and flash units in plastic. I have a gaggle of old Yashica bodies whose lens mounts fortunately accept the Contax lenses without modification. So I just shoot, get drenched, wipe the cmaeras down and shoot some more. I wouldn't recommend this technique to anyone who worries about equipment, but some of my best shots have come from the middle of the action with the much-abused workhorse Yashicas.

What kind of film should I use?

Deciding on film requires an understanding of both the reasons you are taking fire photographs in the first place and the results you hope to produce from your efforts. Will you need transparencies for training slide shows, large blowups for arson evidence, low-priced prints for courtesy handouts to the firefighters or just good sharp images for possible publication in newspapers and magazines?

My own solution is to shoot color slides—transparencies—in most situations because a slide is the only type of original which can be used for anything: reproduction in periodicals, color prints, black and white prints (through use of an internegative) or duplicate slides for audiovisual programs. Slides are easy to file and easy to send out for review. Since I still like to spend time in the darkroom, I usually carry an extra camera loaded with black and white, but when the action is heavy, I prefer color slides.

It is generally agreed that Kodachrome slide film is unbeatable for its sharpness and overall color resolution. Kodachrome comes in two speeds: ASA 25 and ASA 64. Neither is fast enough to be perfect all the time for fire photography, but I still use it whenever I can, even if that means slower shutter speeds, wider lens openings and more powerful flash units. The results are clearly superior.

The higher speed alternative, of course, is to use Ektachrome 200 or 400. This film will result in good images from situations where there is much less light, but the slides will be grainier, often more contrasty and weaker in color vitality. Ektachrome also tends to lend a "colder"—that is, bluer—cast to the pictures. Some interesting new films being produced by Fuji in Japan are similar to the Ektachromes and are processed with the same chemistry, but have much better color fidelity and grain structure.

One great thing about Ektachrome and Fuji films is that private labs can process them in a matter of hours, where Kodachrome, for the most part, must be shipped off to regional Kodak labs. It can often take several days to see Kodachrome slides; even big-city shooters have to wait overnight to see what they've shot. Photographers who absolutely need to have their pictures "yesterday" generally stick with Ektachrome and Fuji.

The new generation of Kodak color negative film is greatly improved. It is also super fast, up to 1000 ASA! All of the color negative films are terrific for fire photography, and those one-hour labs sprouting up around the country offer the fireshooter wallet-size prints in no time at all. The little prints are wonderful handouts for those firefighters who see you shooting and ask for copies.

But I still think that for the important stuff, where you're liable to get only a single quick shot of an exceptional fire scene, you will want that great photograph to be a slide. It can be used for anything from the cover of *Firehouse Magazine* to a stunning color print for the chief's office wall. Experiment with different films, take notes, compare the results and decide which combination best fits your needs.

Should I use strobes to shoot fires?

For many of my shots, electronic flash provides at least a portion of the total illumination. Once again, it is very important to understand the capabilities and limitations of your equipment. The most significant limitation to remember is that even the most powerful strobes can throw their light only a relatively short distance. Forty feet is a long way for any strobe to still be useful, so don't be like the Instamatic-toting teenyboppers who try to strobe their favorite rock star from halfway across a football stadium.

The little rotating scale found on most strobe units will tell you approximately how far you can expect that unit to be effective. These estimates are often optimistic. Take your own unit outside at night, set it up with the camera on a tripod, load the film you normally use and shoot a test roll over carefully measured distances. Bracket your exposures at each distance setting, record all the settings and distances and then examine the processed film to see which frames are perfectly exposed. A roll of calibration film is not wasted, as it will tell you precisely what you can expect from your own electronic flash.

All 35mm cameras have maximum speeds at which the shutter will synchronize with the ultra-fast burst of light from the strobe. Be sure your camera is set correctly: if your shutter speed is too fast only a portion of the film will receive the flash. There is no rule, however, that says you can't use a *slower* shutter speed with your strobe. At night use slow shutter speeds to capture the light from the flames; a blast from the strobe will sharply illuminate the firefighters in the foreground. Try speeds as slow as 1/15th, 1/8th and even 1/4th of a second. Brace the camera on a solid surface or squeeze the shutter gently, trying for a minimum of movement, or use a tripod if you have the time. Even if the background is a bit soft the foreground figures will be motion-free, thanks to the short duration of the light burst from the strobe. Remember to focus carefully on the foreground subjects, using a flashlight to get a fix on them if you have to.

Strobes can also be used as fill lights during the day; many newspaper photographers balance the light with strobe fill on practically every daylight shot. A small shot of light from a small strobe can fill in the shadows below the bill of a fire helmet, for example, adding a nice catchlight in the eyes. Again, monitor the shutter speed to make certain it doesn't exceed the strobe synch speed.

Two final words of caution about strobes and fireshooting. Be very careful about handling the units when they, and you, are wet. Even the little battery models store up an impressive amount of voltage and can deliver devastating shocks. And under the heading of plain common courtesy, be thoughtful when popping your flash units at night, and don't take flash close-ups of faces. The pupils of the eyes are dilated in the darkness, so a flash could easily result in temporary blindness to a firefighter—a dangerous situation to somebody descending a ladder or working at the edge of a roof.

How can I know where the fire action is?

Lots of fire buffs simply follow the call of the siren, hardly a very scientific approach. Your best bet is to buy a scanner, a sophisticated radio receiver that can monitor your local fire frequencies and most other public service channels. Among the various manufacturers, Bearcat probably makes the widest range of scanning radios, from little crystal-controlled units for less than $100 to exotic synthesized receivers that can receive and store thousands of channels. Even the most complex scanner doesn't cost much more than $300.

Your local scanner dealer will tell you which frequencies you'll need to cover. With a small battery-powered unit you will be able to listen on the way to a fire and at the scene. Listening to fire service transmission is legal, though some municipalities and states have laws against car radios that can receive police transmissions. Again, check with your local dealer.

Listening to—and understanding—any public service radio chatter takes hours of practice. In the beginning it all sounds either incomprehensible or boring. Most departments use an array of code phrases and numbers as a shorthand way of describing their activities, and you will need some help from other buffs or scanner owners to become proficient at listening. Most fire alarms amount to

little or nothing, but there are clues which can help you get the jump on the action.

Concentrate on the *tone of voice*. Dispatchers and firefighters in the field are all experienced professionals, but their voices will betray excitement and concern when things are starting to jump. Tone of voice is often the best tipoff to a big burner. I can usually tell within the first three words of the dispatcher's announcement whether we've got a live one or just another routine run with "no further assistance required."

Here's another sure-fire indicator: *telephones ringing in the background*. What that means is lots of citizens are calling the fire department, and all are probably reporting the same event. When you hear those telephone bells, grab your cameras and the portable scanner and head for the door.

How close can I get to the action at a fire?

If you have no official standing as a fire or news photographer, you will usually be kept behind police lines at a big blaze. This is for your own safety: to keep you from being hit by falling debris, choked by toxic smoke or burned by a sudden flashover or back-draft. In addition, you can easily get in the way of working fire-fighters as you are concentrating on getting your shots.

As your fire department becomes more familiar with you and your work, you may be allowed greater freedom of movement at fire scenes. But if you are going to be moving in on the action, it is essential that you also become familiar with firefighting operations. You do not want to hinder the firefighters' primary mission; they have quite enough to worry about without adding you to their list.

Above all, don't become so engrossed in your photography that you lose touch with the dangerous situation around you. There are hundreds of accidents that can befall you at a big job: tripping over hoses or wires, getting stung by downed power cables, sloshing through hazardous chemical runoff, slipping on ice, putting your foot through a burned-out floorboard, stepping on a protruding nail. The list is endless. Firefighters don't get burned very often, but they do get hurt a lot. It is hard to imagine a more dangerous environment.

How should I dress when covering a fire?

There are several schools of thought on this one. If you are going to be closing in on the actual fire scene, especially if you are authorized to be shooting inside the fire lines, you should get permission to dress in full firefighting protective gear: turnout coat, helmet, gloves, steel-toed boots—the works. You should carry markings clearly indicating that you are a photographer, and not a firefighter. Exercise caution—fires are mindless, and a flashover or a collapsing ceiling will not discriminate in your favor simply because you are an unofficial observer.

However, if you are remaining some distance from the fire scene, I think it's best to dress warmly and functionally, but not in a way that attracts attention. I cover fires in some pretty rough neighborhoods—the big burners seem to be concentrated in that part of town—and I'm conscious about staying together with my valuable equipment. Leave the fancy camera bags at home. Carry as little equipment as you can, and stuff extra lenses, strobes, batteries and film into your jacket pockets. Look over your shoulder as often as you look through the viewfinder.

What do firefighters think of amateur fire photographers?

The answer to that one runs the gamut. There are fire photographers (and professional news and TV shooters, for that matter) who make such ridiculous pests of themselves at fires that they wind up being forcibly removed from the scene. Climbing all over the rigs for an unobstructed view, without permission, for example, is a "no no." On the other hand, many departments, because they cannot afford full-time paid photographers, rely on dependable amateur shooters to help them out with public relations, training, fire prevention and even official arson photography. As an amateur fire photographer, you should assess the local situation and approach the firefighters and the department with sensitivity to their particular personalities and needs. If you are good, and if you make your best shots available to the department in a timely and straightforward manner, it is likely your efforts will be appreciated and your relationship strengthened.

How and where can I sell my fire action photographs?

No one gets rich selling fire photography. It is best to do it for the challenge, excitement and fun it offers. However, there are ways to defray some of your expenses. Individual firefighters, like all of us, go bonkers over really sensational shots of themselves in action. Make some good prints—or pay to have them made—and take them around to the appropriate firehouses. Chances are you will see some interest in good color 8x10s. You will also be establishing yourself in the minds of the firefighters as a guy who knows what he's doing at a fire, not just another sideline groupie.

Local newspapers will run good action, usually in black and white. But don't expect big bucks from these sales; the papers consider $25 to $50 to be quite generous. Remember that the news media want prints *right now*. Don't expect to waltz into a newsroom at noon with prints from a 2 A.M. fire and expect the editors to get excited— they are already thinking about tomorrow's news. If you become known as a good fireshooter, particularly if you chase the early-morning blazes, the papers might even agree to process your film for you. Just don't pin too much hope on the dailies. Perhaps the best fringe benefit is the excitement and appreciation you will generate among the firefighters, who always feel their jobs don't get enough accurate media coverage.

Manufacturers of fire equipment and their advertising agencies are definitely interested in looking at top-quality photographs that show their equipment at work on the job. Once again, serious research is needed to develop the leads, but here is where the big bucks are made. Look through magazines like *Firehouse* to see who the fire service advertisers are.

And speaking of *Firehouse,* the editors are always in the market for great fire shots, either for their regular "Hot Shots" department or for entire stories on interesting firefights. Magazines also have deadlines, so if you get good dramatic coverage of a major fire, mail the shots off pronto along with a few typed paragraphs giving the who, what, where, when and why. If you wonder why *Firehouse* never seems to run any shots or stories about your local department, it is probably because *you* haven't been sending them the hot stuff. So get on it!

How can I get really great fire photos?

Although it sounds somewhat facetious, the best way to get great shots is simply *to be there.* That's tougher than it sounds. It often means rolling out of the sack at 4 A.M., hitting the street almost as fast as the firefighters themselves, and driving like crazy to the scene of the fire. Just like the department, when you pull up you have to be prepared: equipment in the trunk, carefully organized and ready for action. The cameras must be loaded, strobe batteries at full charge and bodies wrapped in "baggies" to protect them from water damage. With all but the immense blazes, the best visual part of the firefight is often over in a matter of minutes—usually, just about the time you are piling out of your car. Plan on chasing hundreds of boxes, most of them uneventful, in order to get but a handful of dynamite shots. There are no shortcuts, unless you are a helluva lot luckier than I am.

You know, of course, that flames blowing out of windows aren't the only subjects of great fire photos. Firefighting is about people,

too: fighters, victims, bystanders. If you are too late on the scene for the flashiest part of the fight, look for the smaller moments such as exhausted or injured firefighters, dazed or devastated survivors, ice-encrusted rigs, dramatic patterns of hoses and equipment, even moments of lightheartedness and humor as the battle winds down and everybody starts to relax.

Being in the right place at the right time is unquestionably the key to great fire photography. But, once again, fundamental familiarity with your equipment is equally essential. Know how your camera works in its manual as well as its automatic modes. Check your strobes out on their manual, full-power settings by making measured-distance tests at night. Practice adjusting and loading your camera by feel with your eyes closed; it's often too dark at a fire scene to do it any other way. Dry-fire your camera without film, focusing and changing exposure settings on fast-changing subjects. Practice, practice, practice. Everything happens so quickly at a fire scene, and events are not going to wait while you diddle with your equipment.

How can I show my appreciation to firefighters?

That's easy! Put your photos to work on their behalf, and thereby on your behalf. Do your best to get your hottest shots of fire action into the local papers and other publications. Check in with the department and offer your shots for training, fire prevention and general public relations. If yours is a paid department, introduce yourself to the union leadership; the union has given me its generous support here in Yonkers. Let the firefighters know you are anxious to help them out, then *follow through* when your assistance is solicited. Make no promises you don't intend to keep. If a firefighter asks you for a print of a shot you just took at the fire—and don't worry, he will—follow through and get it to him. He won't forget. And, as a result, your relationship with the people and the department will be strengthened.

How-To

The examples shown on these two pages are intended to help the budding fire photographer add to his bag of tricks. By exercising your own imagination and style, you can adapt these methods to your ideas, to produce satisfying shots that are uniquely your own.

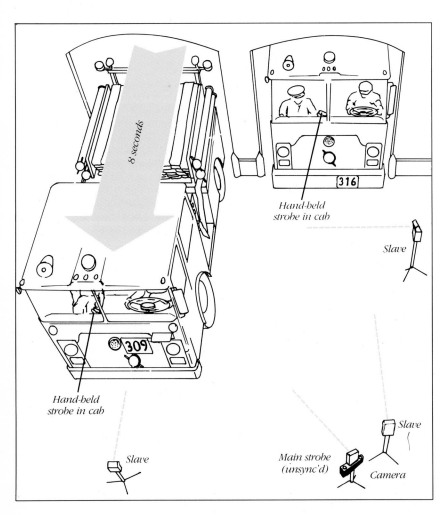

8 seconds

Hand-held strobe in cab

Hand-held strobe in cab

Slave

Slave

Slave

Main strobe (unsync'd)

Camera

Hitting the road. Page *viii*.

Recommended Equipment

Film: Kodachrome, ASA 64.
Lens: 28mm.
Lighting: 2 hand-held strobes, main strobe, 3 slave strobes.

Procedure:

1. Hose down driveway.
2. Driver starts rig moving slowly.
3. Open shutter on B setting, with aperture set at f:8.
4. Rig should be stopped after 8 second exposure.
5. When rig stops, fire main strobe—this causes slaves to fire, as well.
6. This is signal to officers in cab to fire hand-held strobes.
7. Close shutter when hand-helds discharge.
8. Approximate time lapse from time rig starts rolling to shutter closing should be 10 seconds.

Remarks

Beware of bumps in the driveway and short stops by rig. Either of these will cause disruption in the nice clean lines of streaking lights.

One in a million. Page 3.

Recommended Equipment

Film: Kodachrome, ASA 64.
Lens: 50mm.
Lighting: 13,000 volt strobe, generously provided by local power utility.

Remarks

This shot provides a valuable, though exaggerated lesson. Of course, you may spend a month of Sundays and still not encounter such an extreme example of available light, but it does point out the pitfalls of strobing at the wrong time.

My first inclination was to bombard this scene with my flash, but the powerful light provided by the exploding transformer made this a case of "gilding the lily." The result of my restraint was the striking steel-blue light in the mid- and background, with no fill-lighting in the foreground.

Had I used a flash, all sorts of distractions would have raised their ugly heads, and the impact of the real action would have been diminished.

Other, less dramatic available light shots (such as the one on page *x*) come along with surprising frequency. When they do, resist the urge to "dial x." You will get some pleasant surprises when you get that roll of film back from the lab.

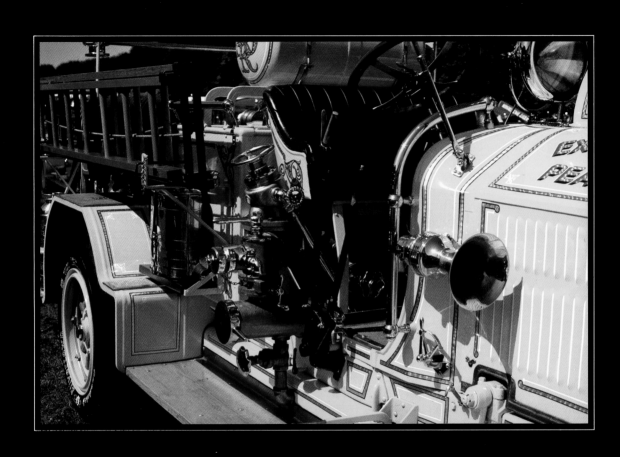